"Production & Processing of Lemongrass Oil"

AUTHORED BY

Govind Ram Jat
Associate Lecturer
Department of Chemical Engineering
Mewar University, Chittaurgarh Rajasthan -312901

DEDICATED TO

My beloved Parents **(Chandi Devi - Prabhu Lal Jat)**,

My lovely Wife **(Neeta Govind Ram Choudhary),**

My beloved Brothers (**Arjun Lal Jat- Satyanarayan Jat)**,

My beloved **Sisters** (**Kamla Devi-Leela Devi**)

&

Chairman of Mewar University**, Dr. Ashok Kumar Gadiya**

ACKNOWLEDGMENTS

We wish to express sincerely in great esteem, our gratitude and thanks to the Almighty **GOD** for providing us through all the channels to achieve this stage.

We owe our regards to our project guide **Mr. Gedam Vilas Dekhelji** (Project Coordinator & guide) & **Mr. S. V. Shukla, Principal Director FFDC, Kannauj** for their generous support to do a Project. It was only due to their painstaking efforts and help at various stages that the presentation of the work has been possible.

I extend my sincere thanks to **my beloved Wife Mrs. Neeta Govind Ram Choudhary** & my friends who are there with me directly or in directly during theWork.

I want to greet **Dr. Ashok Kumar Gadiya, (Chairperson, Mewar University)** for sending me to such type of great MSME institute to complete my Project Work.

We would like to express sincere thanks to **Pratibha Choudhary(beloved sister in law), my parents (Mrs. Chandi Devi-Mr. Prabhu Lal Jat),** for enlightening the spirit of dedication in us.

Govind Ram Jat

ABSTRACT

Keywords:-

Fractionation, Reflux Ratio, Extraction, Cymbopogon citrates, Thermic fluid, Distillate.

The aim of this work was to study the feasiblity of simultaneous process & fractionation of Lemongrass Oil by packed fractionation column. As per Requirement the high quality vacuum (5 mmHg) & low temperature (84 $^{\circ}$C) was applied in the column/system to fractionate the Citral (major part) at low temperature in the fractionation column.

Also the purpose of this work was to apply the material balance on the fractionation column & estimate the process loss of the material in the packed fractionation column.

As per the process economics of the fractionation process, we found the total margin (profit) of the process is almost **165 INR / kg.**

The monoterpene (which is first fractionate) also separated at low temperature From the column and used in the process economics of the fractionation process.

The energy balance also applied on the condensation of the vapor of the product citral by the energy balance principal of the condenser.

Contents

ABSTRACT 4
1.0   INTRODUCTION 7
2.0   HISTORY 9
3.0   BOTANY 10
    i.   TAXONOMY 10
    ii.   PLANT DESCRIPTION 10
4.0   HABITAT 11
5.0   AGRONOMY 12
    i.   SOIL AND CLIMATE 12
    ii.   PROPAGATION 13
    iii.   FERTILIZER 14
    iv.   WEEDING AND INTER-CULTURE 15
    v.   HARVEST AND YIELD 15
    vi.   DISEASES AND PESTS 15
6.0   CHEMISTRY OF LEMONGRASS 17
    i.   PHYSICO-CHEMICAL CHARACTERISTICS 17
    ii.   BIOCHARACTERISTICS 19
7.0   EXTRACTION OF LEMONGRASS OIL BY LEMONGRASS 21
    i.   DRYING 21
    ii.   DISTILLATION 21
        STEAM DISTILLATION 21
    iii.   PURIFICATION OF OIL 28
8.0   STORAGE AND PACKING OF OIL 28
9.0   FRACTIONATION OF LEMONGRASS OIL 28
    i.   FRACTIONATION AT LAB SCALE 29
    ii.   Fractionation at Industrial Scale 30
        Process Description 30
        Strategy of Fractionation Column Startups and Shutdowns 31
10.0 TROUBLESHOOTING 41
    i.   Technique for Preventing Tray Damage Due to Excessive Liquid Level 41
    ii.   Sources and Effects of Water Problems 41

11.0	Lemongrass Oil Processing Economics	42
12.0	Uses of Lemongrass Oil	43
i.	PERFUMERY	43
ii.	INSECT REPELLENT PROPERTIES	43
iii.	GERMICIDAL AND BACTERICID	44
iv.	THERAPEUTIC USAGE	44
v.	OTHER USES	44
13.0	REFERENCES	45

List of Figure

Figure 1 Lemongrass Plant at FFDC Kannauj (U.P.) 7
Figure 2 Forced Draft Cooling Tower 20
Figure 3 Induced Draft Cooling Tower 21
Figure 4 Cooling tower at FFDC Kannauj Utter Pradesh 21
Figure 5 Fire Tube Boiler 22
Figure 6 Water Tube Boiler 23
Figure 7 Simple Distillation Plant 24
Figure 8 Diagram of Fractionation Column with Reflux and Distillate 27
Figure 10 Packed Distillation Column 31
Figure 11 Material Balance on Fractionation Column 33
Figure 12 Glass Fractionation Unit (Capacity-25 liter) in Process Division at FFDC, Kannauj 36
Figure 13 Energy Balance on Condenser of Fractionation Column 36

1.0 INTRODUCTION

Lemongrass is an aromatic grass belonging to the family Gramineae and genus Cymbopogon, which consist of about 80 species. Lemongrass is well known for its Oil and it is one of the world's best known essential oils. There are two main types of lemongrass namely East Indian and West Indian.

The East Indian lemongrass oil is obtained from Cymbopogon flexuosus Stapf and is the genuine oil of commercial importance. The species is considered to have originated in Kerala, the southern most state of India. According to the colour of the stem, this is again divided into two types. The 'red grass' which is true Cymbopogon flexuosus is known as 'choomanna poolu' in Tamil; and 'white grass' known as 'wella poolu' has been identified as Cymbopogon flexuosus var. albescens. The oil obtained from this plant has very low aldehyde content and is poor in solubility.

The West Indian oil is extracted from Cymbopogon citratus (DC) Stapf that is mainly cultivated in Central and South America and also known in parts of Africa, South East Asia and the Indian Ocean Islands.

The name lemongrass has been given because of its typical strong lemon - like odour, which it due to the high citral content. The two oils were formerly the main source of natural citral. The essential oils are used in perfumery, cosmetic and pharmaceutical preparations. However, lemon-grass oil has declined in commercial importance due to the competitive synthesis of citral and isolation of natural citral from Litsea cubeba oil. Nevertheless, more than 1000 t/a are still produced.

A third species, Cympobogon pendulus (Nees ex Steud) Wats has been recently distilled in India to a limited extent. This grass also contains 70 - 80% citral. However it is much harder than the two previous species and can be cultivated in adverse soil and climatic conditions. It is popularly known as Jammu Lemongrass or RRL-16.

2.0 HISTORY

Lemongrass has been used in medicine in India for more than 2000 years. However, its first recorded distillation is in Philippines in the 17th century, and in 1799 it was introduced to Jamaica. In 1917 it was grown for the first time in USA by Hood and during the World War I, grown in Guatamala by Julia Samayoa. Although the oil has been known, since very early times in India, the systematic cultivation and distillation of the grass were started in Kerala only about 100 years ago. Cultivation has assumed the status of a plantation crop after World War II.

Figure 1 Lemongrass Plant at FFDC Kannauj (U.P.)

3.0 BOTANY

i. TAXONOMY

Family: Gramineae

Genus: Cymbopgon

Species: citratus, flexuosus

Botanical Names

Lemongrass, West Indian type: - Cymbopogon citratus (D.C.) (Stapf); Andropogon nardus var. ceriferus (Hack); Andropogon citratus (DC.)

Lemongrass, East Indian type: - Cymbopogon flexuosus (D.C.)(Stapf); Andropogon flexuosus (Nees)

Other names:-

Sinhala	:	Sereh
English	:	Lemongrass
Tamil	:	Wei la poolu, Ondapoolu (White grass), Choomana poolu (Red grass)
Sanskrit	:	Bhustrina, Takratani.

ii. PLANT DESCRIPTION

Cymbopogon flexuosus:-

About 3 m tall grass arising from a woody rhizome. Leaf sheath glabrous, hairy at the junction blade. .Leaf blade 1 m long, 1.5 cm wide, linear, acuminate, glaucous. Panicles are very large, drooping, lax, greyish or greyish green, rarely with a purple tinge, with raceme pairs in dense masses, spreading, 100 - 135 cm tall, slightly hairy, lower glumes of the sessile spilelets 3-4, rarely 4 - 5 mm long, 1mm wide with 1-3 definite or obscure introcrainal nerves, shallowly concave with one or two depressions.

Cymbopogon citratus:-

It is a perennial aromatic grass having dense fasicles of leaves from a short/oblique annulate, sparingly branched rhizome. Leaf blade is linear, long - attenuated towards the base and tapering upwards, approx. 90 cm long, 5 cm wide, smooth or rough upwards and along the margins, glabrous, glaucus green, base narrow. Sheaths are terete, those of the barren shoots much widened at the base. Inflorescence spatheate panicle, decompound to sub-decompound, lose, 30 - 60 cm long, internodes 4 to over 6, spatheoles very narrow, linear lanceolate to almost subulate when INRolled. Peduncles 6-10 mm long, glabrous. Fertile spikelets linear, lanceolate,' acuminate, 5-6 mm long, reddish glaborus, lower floral glume hyaline, linear, oblong or almost linear.

Cymbopogon pendulus:-

It is a perennial robust grass. Clump is erect, 120 - 150 cm high and glabrous. Leaf- blades are upto 80 cm long, 11 mm wide. Slightly rough on the low surface, glabrous, leaf-sheaths tomentose below. Spathate is panicle, 60 cm long, more or less interrupted, each tier composed of three rays in cluster spatheole 15 to 26 mm long; rachis internodes and pedicels pilose along the margins and on the upper part. Lower glume is yellowish, lanceolate to ovate lanceolate, 1-1.2 mm wide; awn 11 mm long. Pedicellate spikelet, 3.8 mm long. Lower glume is 0.8 mm wide.

4.0   HABITAT

Lemongrass grows wild in India and other tropical areas. Cymbopogon flexuosus is indigenous to India and is cultivated in Ethiopia, Guatemala, Indonesia, Japan, Madagascar, Seychelles, Sri Lanka and Thailand. Cymbopogon citratus is found in Argentina, Brazil, Cameroon, Cuba, Gautamala, Haiti, Indonesia, Jamaica, Japan, Kenya, Mexico, Philippines, Seychelles, Somalia, Surinam, Tanzania, Thailand, Uganda and Zaire.

5.0 AGRONOMY

i. SOIL AND CLIMATE

Lemongrass grows in tropical climate. It grows on poor soil and can be planted in areas where citronella does not thrive. It is harder than citronella and more resistant to drought.

Elevation:-

The plant grows best at elevations ranging from sea level to 1200m.

Rainfall:-

The grass prefers an annual rainfall of 200 - 250 cm. In regions of abundant rainfall, the plant may be harvested more frequently during the year, but the oil will have lower citral content. Prolonged rain is harmful. Planting on ridges is recommended in these areas.

Temperature:-

The temperatures most suited for growth of this grass is from 24 - 29 °C.

Climate:-

A warm and humid climate with plenty of sunshine is necessary for growth. Sunshine is said to influence the citral content of the oil.

Soil:-

The grass is found to be good soil binders and also act as a vegetation cover over naked eroded slopes in the foothills but none of these can stand water – logged condition. The plant flourishes on a wide variety of soils ranging from rich loam to poor laterite. The red yellow podsolic soil is equally good.

Higher soil pH (7.5) significantly increases the yield of herb and oil content, total citral and citral 8 as compared to lower pH (4.8).

ii. PROPAGATION

Lemongrass can be grown as a catch crop for rubber. It grows in rows between trees, without interfering with rubber roots. It has also been planted in between coffee trees.

Cymbopogon flexuosus:-

The plant is propagated both from seeds and rooted slips. Propagation through vegetative means from selected clones is considered better as seed propagation leads to considerable genetic heterogeneity resulting in deterioration of yield and oil quality.

Seeds are the best means for rapid and large scale cultivation of East Indian lemongrass. Seeds are collected from plants, which are not subjected to regular harvest. The whole inflorescence is cut, dried for 1 to 2 days, thrashed and the seeds separated. In direct seeding, the land is ploughed and seeds sown in March at the commencement of the rainy season.

For raising a nursery, the soil is ploughed 6 to 8 times before sowing of seeds. The seeds are sown between April and May. The seedlings produced by 10 to 12 kg of seeds will be sufficient for transplanting into an area of one hectare.

Seeds start to germinate from the 3rd day of sowing and continue for about one month. More than 90% germination was completed by the first week itself. Hence it could be assumed that the seeds which did not germinate within one week may not be germinated later. In such cases, fresh seeds could be sown immediately to save time. .

The seeds sown immediately after collecting (without storage) exhibited a comparatively low germination. After a storage period of 60 days, germination was increased suddenly.

The seedlings are transplanted when about 60 days old. The tips of seedlings are cut leaving the planting material 15 to 20 cm long. They are then planted along ridges of about 90 cm, in holes about 10 - 12 cm apart. Two or three seedlings are planted in the same hole. A spacing of 45 X 45 cm has been reported to give high yields of oil containing a high percentage of citral.

Seeds of Cymbopogon flexuosus var. OD 19, subjected to a - irradiation at a dose of 20 krad resulted in significant increase in herbage and oil yield whereas significant increase in citral was achieved by 10 krad dose.

Cymbopogon citratus:-

Cymbopogon citratus is propagated by means of root divisions at the onset of the monsoons. The clumps of healthy mature plants are divided carefully into a number of slips, each slip containing 1 to 3 tillers. The tops and fibrous roots of the slips are trimmed off before planting. One old clump could yield as much as 50 new stools.

The soil must be loosened thoroughly by ploughing before planting. Holes about 20Cm deep are dug with a crowbar and the slips are planted 45 - 60 cm apart, in rows about 90 cm apart. One to four segments are planted in each hole depending on the fertility of soil. The planting holes are lightly filled with earth to facilitate the development of a good root system. A 100 sq.m nursery would yield sufficient root segments to plant 1 ha, with two root segments per planting hole. Thus 15,000 to 20,000 segments can be planted/ha.

In a study it has found that it can be grown effectively in hydroponic in open. A mixture of gravel volcanic slag or gravel and pumice yields herbage of 55 tonnes/hectare and essential oil up to 204 kg/hectare.

Cymbopogon pendulus:-

The crop is propagated vegetatively through slips obtained by the splitting up of individual clumps which give about 110 - 150 tillers per clump. Clumps bearing well over 200 slips have also been observed. The crop is planted on flat beds and irrigated immediately afterwards. One to two healthy slips about 20 cm in length are planted per hold. The crop establishes itself within 30 days and thereafter tillering starts.

Results over a period of 5 years have shown that narrow spacings in between plant rows give high yield of herb and oil. Spacing between rows should be kept at 50 cm, while plant distance may vary from 30 - 50 cm.

iii. FERTILIZER

Lemongrass is not frequently manured as it is known to be a soil exhausting grass. However the application of fertilizer when the plants have become well established is said to be beneficial. It is recommended that in soils of average fertility, a mixture of 30 kg/ha N, 30 kg/ha of P and 30 kg/ha of potash is applied at the time of planting. 60 kg/ha of N is applied as top dressing in 2-3 split doses during the growth season. Fertilizer also can be applied by introducing into holes in which the grass is to be planted and covering with 3 to 6 cm. of soil. It could also be broadcast, hold or disked. Better growth has been obtained when potash was supplied as the sulphate and when part of N was organic.

The micronutrients, Bo and Cu significantly increase the grass yield as well as growth. The copper gives the maximum grass yield, while application of other micronutrients, such as Mn, Zn, Mo, Si had no significant influence on oil yield.

Eastern India:-

A mixture of 60 kg/ha of N, 55 kg/ha P_2O_5 and 30 kg/ha of K_2O has been found to be effective.

West Bengal:-

N: P: K: 60:50:35 kg/ha was most effective, and induced maximum vegetative growth, dry weight, accumulation and essential oil format.

North India:-

To get optimum yield, an adequate mixture of NPK is required to meet the needs of the crop. The crop has high requirement of nitrogen and at least 250 kg/ha should be used for economic returns. A dose of 80 kg $P2O5$ and up to 120 kg of $K2O$ per ha is applied at planting or at the time of first weeding. Application of N in split doses (3-4) is recommended.

iv. WEEDING AND INTER-CULTURE

Weeding and hoeing are very important for the yield and quality of oil. Generally 2-3 weedings are necessary during the year. Earthling up should also be done at least once in a year. During the winter, the grass is usually burnt to fertilize the soil and to strengthen the plants. In row-planted crop, inter-culturing can be done by a tractor- drawn cultivator or hand-hoe. Weeds can also be controlled through the application of oxyflureofen, diuron and simazine at the rate of 0.5, 1.5 and 2.0 kg ai/ha, respectively. Distillation waste, when applied at 3 to 5 tonnes/ha suppress weed growth and is equally effective.

v. HARVEST AND YIELD

The time for the first harvest varies from 3-9 months and subsequently the grass may be harvested every 3 or 4 months. When the grass is about 4 ft. high and has 4 – 5 leaves, it is ready for cutting. The higher yields of leaves and oil are obtained from the second to the sixth year, after which there is a steady decrease.

Harvesting is done by sickle by cutting the leaves 10 - 15 cm above the ground level. Plants grow rapidly after each cutting. The number of harvests depends on the climatic conditions of the place of cultivation. Time of harvest is important for good yield of oil and its quality. It was found that maximum oil yield for Cymbopogon flexuosus is obtained when harvested at a maximum height and the oil content was higher when harvested during night than that harvested during the daytime. It is lowest when harvested between 12.00 noon to 4.00 p.m. Cymbopogon pendulus should be harvested before flowering. Flowering adversely affects the yield and quality of oil and is reported to reduce the oil yield by 30%.

The percentage recovery of oil was greatest in grass harvested every 30-35 days and declined thereafter. However, the citral content of the oil increased with increasing time intervals. Immature leaves were found to possess the greatest amount of essential oil and percentage of citral, with respect to the leaf blade.

vi. DISEASES AND PESTS

Lemongrass is subject to attack by many pest and diseases.

DISEASES:-

Leaf rust disease

Prolong rain causes attack by rust. This causes serious losses to herb and oil yield.

Eye spot disease

Caused by Helminthosphorium organism, which attacks sugar cane. In this disease the spots appear at first as minute yellow flecks with rusty brown centre. The spot grows an oval portion and becomes a light flesh or straw colour, outside of which is, a purple, narrow border surrounded by a oval zone, Spanish - raisin in colour. This is surrounded by a yellowish to flesh-coloured aureolas. Often oval spots coalesce to form irregular blocks.

Chlorosis

Iron chlorosis is due to deficiency of iron leading to loss of green colour of leaves. This results a reduction in yield of oil and concentration of major compounds. Iron chlorosis is seen, when irrigated with water containing bicarbonates. or when grown on calcareous soils.

This could be controlled by foliar application of 3% ferrous sulphate.

Attack by Fungus

Foliage is susceptible to attack by the fungal pathogen.

A .The leaf tips and margins are attacked by the fungus **Curvularia andropogonis** causing 10 - 20% loss in green tissue. Infected tissue turns brown and eventually becomes necrotic, giving the appearance of scorched tips.

Controlled by using fungicides captan, zinneb, ziram.

B. Grassy Shoot Disease:

It is prevalent in several parts of India, especially in Kerala. The disease is caused by the fungus Balansia sclerotica. This is caused by excessive profileration of shoots and shortening of leaves and conversion of flowering parts into leafy structures giving the typical appearance of little leaf or witches broom type of disease.

C. Leaf blight:

Several fungi cause it, but the most severe infection is caused by Colletotrichum graminicola. This disease appears in the form of small brownish spots on the .leaf stem and enlarges into large brown patches. It has also been found that leaf blight can also be caused by Rhizoctonia solani.

D. Smut:

It is caused by the fungus, Tolyposporium christensenii. The symptoms appear in the form of conversion of flowering panicle into black powdery mass. Seeds and flowers are converted into smut spots.

INSECTS AND PESTS:-

The most destructive pest infecting lemongrass is a species of Chilotrea. The caterpillar is white in colour with a black head and black spots on the body. It bores into the stem and remains there feeding on the shoot. It is usually found at the bottom of the stem.

The first symptom of the attack is the drying up of the central leaf. Subsequently, the whole shoot dies, resulting in a significant reduction in the yield of the grass.

Control Measures:-

(i) The dry stubble is set on fire during the off season in summer. The caterpillars lurking inside the stubble are thus destroyed.

(ii) The affected shoots are pulled out and destroyed.

(iii) When attacks are serious, a spray of oxydemeton-methyl is used. Instructions for spray, etc. should be followed as prescribed-by the manufactures.

Sugar cane borer, Sesamia inferens and the Madagascar beetle, Heteronychus plebejiis attack freshly planted lemongrass. Scale insect, Duplachionapsis divergens produces yellow spots on the stem and sucks the sap of the leaves and stem. White fly, Tetraleurodes semiletmanaria sucks sap from the leaves causing chlorosis and withering away of leaves. The insect can be controlled by 0.5% Dimethoate.

Clovia bipmctata, a spotted bug, attacks the lemon grass in Kerala. The nymphs attack young leaves causing crinkling. It can be controlled by 0.05% Quinolphos.

Lemongrass is also attacked by the aphid, Sipha flava, which causes the grass to become reddish brown and dry. Although the yield of oil from infested grass is somewhat lower, the infestation does not markedly affect the quality of the oil.

Roots are attacked by nematodes. Infested clumps turn yellow, wilt and wither away.

6.0 CHEMISTRY OF LEMONGRASS

i. PHYSICO-CHEMICAL CHARACTERISTICS

A).ESSENTIAL OIL:-

The oil has an intense lemon - like odour and taste, and is yellow, reddish-yellow to

reddish brown in colour. West Indian oil has an odour similar to that of Citronella.

Physical constant	Cymbopogon citratus
Specific gravity 20°C	0.8986
Refractive index 20°C	1.4910
Optical rotation 20°C	-0.62
Acid value	5.34
Ester value	44.20
Carbonyls	74.96%

B). Moisture content:- One year- old oil was found to have about 0.1% moisture. The oil can retain 2.5 -3% dissolved water at room temperature. Water when present in the oil is harmful to citral. In the presence of water, air and light, citral decomposes rapidly.

C). Solubility:-

The West Indian oil is usually less soluble in alcohol than the East Indian oil. The lower solubility in alcohol was at one time attributed to the steam distillation method, higher boiling and less soluble products being carried over during steam distillation. The lower solubility has also been explained as due to the presence of myrcene, which on exposure to air and light readily polymerizes. According to the ISO standards solubility of the oil decreases on storage. The oil from white grass shows poor solubility.

D). Spent Grass:-

Proximate analysis {Cymbopogon citratus}

The dust passes through and is retained on 80 mesh.

Ash	6.2%
Solubility (cold water)	12.0
Solubility (hot water)	21.4
Solubility (1% caustic soda)	45.6
Solubility (Ale. Benzene)	9.2
Pentosans	16.6%
Lignin	21.3%

Cellulose   46.9%

E). Structure of Citral

Citral

ii. BIOCHARACTERISTICS

A). Anti-bacterial properties:-

Bacterial efficiency of lemongrass oil was found to be proportional to the citral content of the oil. Mycene enhanced the activity of citral when combined with them. Tests on the influence of different emulsifiers on the bactericidal action of the oil showed that of the emulsifiers, triethanolamine or potassium oleate and rosin soap were the best. No relationship has been observed between viscosity of the emulsion and bactericidal efficiency. Excess solvent lowered stability of the emulsion and also lowered bactericidal action of the emulsion.

The essential oil of Cymbopogon citratus shows rapid bacterial activity against both Gram-negative and Gram-positive bacteria in laboratory studies. Gram-positive bacteria were the more active bacteria.

Cymbopogon citratus oil has shown detectable activity against Bacillus subtilis, Staphylococcus aureus and Escherichia coli. But when the oil was oxidized, activity reduced and was completely lost when the oil was extensively oxidized. Inclusion of antioxidants in the oil samples reduced the ratio of oxidation and enhanced the antibacterial activity.

The combined use of lemongrass oil (LGO) and phenoxyethanol can increase the spectrum of activity of phenoxyethanol whose activity is mainly against Pseudomonas aeruginosa. An appreciable increase in the activity of phenoxyethanol against Escherichia coli and Staphylococcus aureus has been observed when it was combined with 0.03 v/v LGO. In addition, the combination reduced the effective concentration of both components necessary for activity.

B). Anti-fungal properties:-

The anti-fungal properties of lemongrass oil and its fractional constituent, citral A, have inhibited the growth of Aspergillus niger at 0.20% and 0.15% concentration respectively in agar medium. The pH of the medium was found to have no effect on inhibition of the growth of the mold. Inhibitory action of lemongrass oil is considered mainly due to its citral A content.

Lemongrass oil a shown good anti-fungal activity against Aspergillus species: A. niger, A. flavus, A. Jumigatus, A. nidulans and Fusarium oxysparum at a concentration of 2.5 ml of oil in 1000 ml of medium.

The essential oil of Cymbopogon flexuosus from Pakistan and Thai cultivars showed inhibitory effect against pathogenic fungi, Monilia sitophilia, Penicillum digclatum, Aspergillus parasiticus, Aspergillus niger and Aspergillus fungis.

Steam distillate from Cymbopogon citratus completely inhibit the growth of Ustilago maydis, Ustilaginoidea virens, Curvularia lunata and Rhizopus species.

Application of lemongrass oil as preservative to control Blue Mould Decay by Penicillium italicum, in oranges is shown to be effective and dosage comparable with those chemicals used in United States.

C). Anti-viral activity:-

Essential oil of Cymbopogon citratus was found to be potent in reducing Potato virus X and Potato virus Y by 100% up to a dilution of 4:1000.

D). Insecticidal and pest repellent properties:-

Lemongrass repels the tsetse fly. Creams containing the oil are used as repellents against houseflies and mosquitoes. Contact toxicity of the oil in acetone solution on insects is very high. 7.5 µg per insect was effective in killing houseflies and 30 fig per insect killed mosquitoes.

In an experiment to assess the toxicity of oil extract from Cymbopogon citratus against Sitophilus zeamais, it was found that there is an increase in mortality of this pest compared with the control at dose rates of 0.1, 0.5, 0.7, & 1.0 ml oil/50 g maize.

Cymbopogon citratus also exhibited high repellency to Dacus dorsalis. Root extract of Cymbopogon citratus, when compared with fenamiphos (15 kg/fed) for control of Meloidogyne javanica on tomatoes in a pot experiment showed 55.4% reduction in nematode population and 57.1% reduction in egg masses 60 days after inoculation.

Polyolefin-paper packing laminates when treated on the paper side with lemongrass oil showed effective dog and cat repelling properties even after 7 days. Extract of Cymbopogon citratus had shown significant anti-feedant and insecticidal effects on 3rd instar larvea of Crocidolomia binotalis, one of the most important pests of cruciferous crops in Mauritius.

7.0 EXTRACTION OF LEMONGRASS OIL BY LEMONGRASS

I. Drying
II. Distillation
III. Purification of Oil

i. DRYING

The grass is allowed to wilt for 24 hours before distillation as it reduces the moisture content by 30% and improves oil yield. The crop is chopped into small pieces before filling in the stills. It can be distilled in same distilleries as used for Japanese mint in India.

ii. DISTILLATION

Oil is obtained through steam distillation. The oil has a strong lemon like odour. The oil is yellowish in colour having 75-85% citral and small amount of other minor aroma compounds. The recovery of oil from the grass ranges from 0.5 - 0.8 per cent. It takes about 4 hours for complete recovery of the oil.

STEAM DISTILLATION

The main components of steam distillation unit are:

1. Cooling Tower

2. Boiler (Steam Generator)

3 Distillation tank with Direct steam & Jacketed Steam

4. Condenser (usually multi-tube tubular)

5. Oil separator or receiver

COOLING TOWER:-

A cooling tower is a simple device used by industry to remove heat from water. Hot water transfers heat to cooler air as it passes through the internal components of the tower. This type of heat is called sensible heat; sensible heat can be measured or felt. Sensible heat accounts for only 10% to 20% of the heat transfer in a cooling tower. Most of the heat stripping from a tower is caused by evaporation. Evaporation accounts for 80% to 90% of the heat transfer in a cooling tower. When water changes to vapor, it takes heat energy with it, leaving behind the cooler liquid. The principle of evaporation is the most critical factor in cooling-tower efficiency.

A cooling tower is a large rectangular or box-shaped device filled with wooden or plastic slats and louvers that direct airflow and break up water as it falls from the top of the water distribution header. The internal design of the tower ensures good air and water contact.

Cooling towers are classified by (1) how they produce airflow, and (2) the direction the airflow takes in relation to the downward flow of water. Airflow may be produced naturally or mechanically. Mechanical drafts are created by fans located on the side or top of the cooling tower. Flow direction into a tower is either cross flow or counter-flow. Cross flow goes horizontally across the downward flow of water before exiting the system. When the air is forced to move vertically upward, against the downward flow of water, it is referred to as counter-flow. Cooling towers come in the following designs:

Natural Drafts:-

- Atmospheric—simple counter-flow
- Hyperbolic (chimney towers)—counter-flow or cross-flow

Mechanical Drafts:-

- Forced draft—counter-flow
- Induced draft—counter-flow or cross-flow

The basic components of a cooling tower include a water basin, pump, and water make-up system at the base of the cooling tower. The internal frame is made of pressure-treated wood or plastic and is designed to support the internal components of the tower. Some of these components include the fill or splash boards and drift eliminators. The fill or splash boards enhance liquid air contact, while the drift eliminators reduce the amount of water lost from the tower because of excess airflow. Louvers on the side of the cooling water tower admit air into the device. A hot-water distribution system is typically located on the top of the cooling tower fill. A fan may be used to enhance airflow through the cooling tower. Fan location determines whether airflow is induced (drawn in) or forced (pushed in).

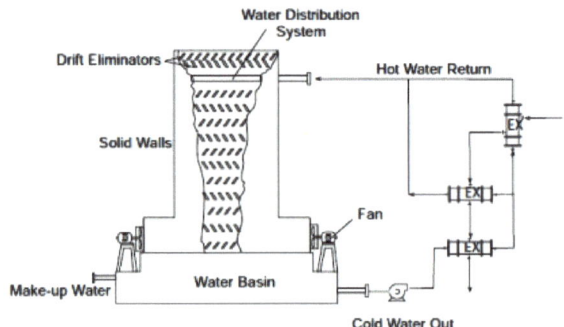

Figure 2 Forced Draft Cooling Tower

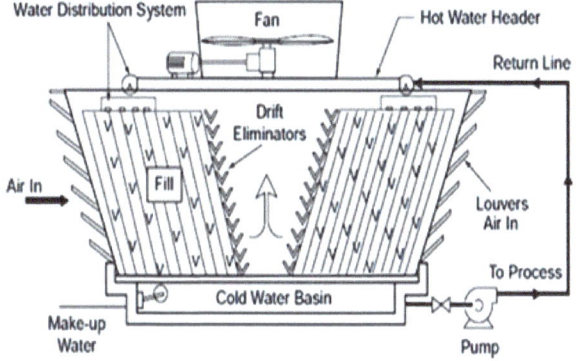

Figure 3 Induced Draft Cooling Tower

Cooling Tower of Process Plant at FFDC, Kannauj (U.P.)

Figure 4 Cooling tower at FFDC Kannauj Utter Pradesh

BOILER (STEAM GENERATOR):-

Steam generators, commonly called boilers, are used by industrial manufacturers to produce steam. Steam is used to drive turbines and provide heat to process equipment. Steam generators are classified as fire-tube or water-tube boilers. High-pressure, medium-pressure, and low-pressure steam is circulated and used in numerous applications within a typical plant. Water-tube boilers are typically designed for large industrial applications; fire-tube boilers are used in smaller systems and processes.

Fire-Tube Boilers

Fire-tube boilers contain the combustion gases in tubes that occupy a small percentage of the overall volume of the heater. The heated tubes run through a shell that contains the heated medium (water). A fire-tube boiler resembles a multi-pass shell-and-tube heat exchanger. This type of boiler is composed of a shell and a series of steel tubes designed to transfer heat through the combustion chamber (tube) into the horizontal fire tubes. Exhaust fumes exit through a chamber similar to an exchanger head and pass safely out of the boiler. The water level in the boiler shell is maintained above the tubes to protect them from overheating. The term fire tube comes from the way the boiler is constructed.

The basic components of a fire-tube boiler include a large shell that surrounds a horizontal series of tubes. A large, lower combustion tube is attached to a burner that admits heat into the tubes. The upper tubes transfer hot combustion gases through the system and out the stack. Airflow is closely controlled with the inlet air louvers and the stack damper. Water level in the shell is maintained slightly above the tubes. As heat energy is transferred into the water, the temperature rises until the fluid boils. A pressure control valve maintains the correct operating pressure on the vessel. Every fire-tube boiler is equipped with a pressure relief system. A series of safety valves may be located on the discharge side of the shell. Low-pressure steam is discharged into a common steam header that is connected to various locations in the facility. A condensate return line admits the condensed steam into a deaerator drum and the water make-up system.

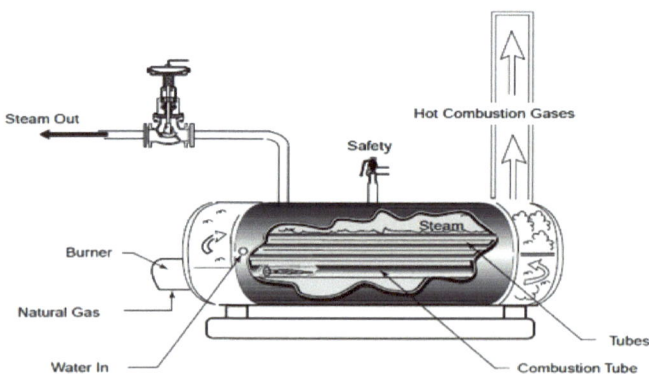

Figure 5 Fire Tube Boiler

Water-Tube Boilers

The chemical processing industry also uses large industrial boilers commonly called water-tube boilers. A water-tube boiler consists of an upper steam-generating drum and a lower mud drum connected by three types of tubes: down comers, risers, and steam-generating tubes. These drums and tubes are surrounded by a furnace and a series of specially designed burners. The lower mud drum and water tubes are completely filled with water, whereas the upper steam-generating drum is only partially full. This vapor cavity allows steam pressure to build, collect, and pass out of the header. Water is carried through tubes that flow near and around the burners. As heat is applied to the water-generating tubes and drums, the water circulates around the boiler, down the down comer tube, into the lower drum, and back up the riser tube and steam- generating tubes of the furnace. During normal operation, high-pressure steam is superheated and sent to the main steam header. Lost water in the boiler is replaced by the make-up water line.

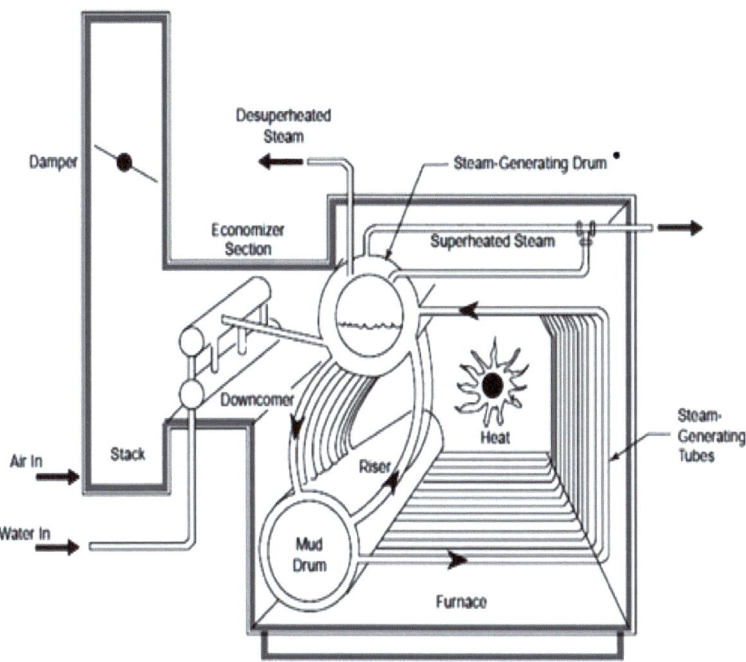

Figure 6 Water Tube Boiler

Distillation Tank with Direct Steam & Jacketed Steam, Condenser and Oil Separator or Receiver

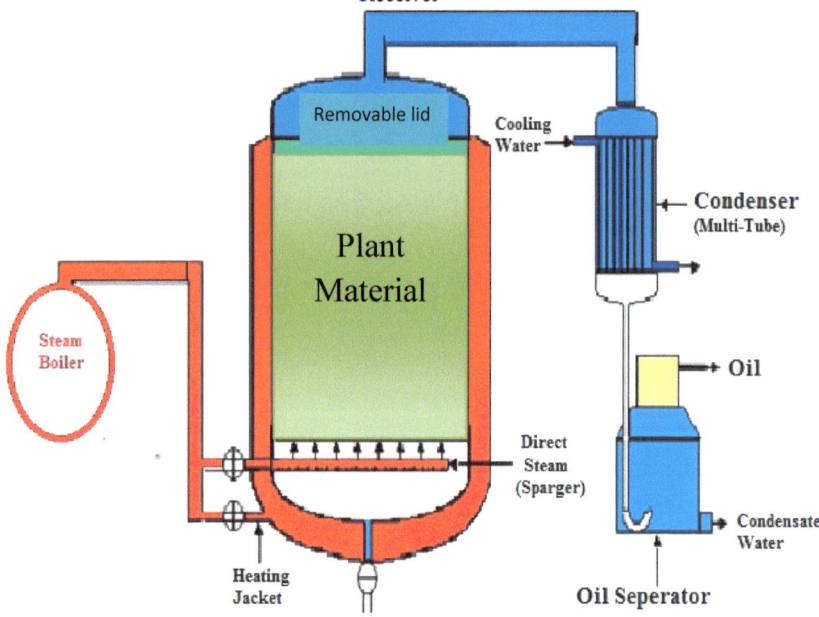

Figure 7 Simple Distillation Plant

In this method steam pressure is injected in boiler which is higher than atmospheric pressure. During the distillation the temperature of the charge is rises to the temperature of boiling water. The superheated steam has tendency to dry out the charge and reduce the rate of recovery of essential oil. Hi pressure steam causes considerable decomposition .This method is better than the water distillation water and steam distillation methods as regards to cost, rate of distillation, capacity of production and the quality of distilled oil. In particular cases the high temperature of the steam destroys the perfume, the amount of oil is relatively small. Steam distillation method is not suitable for flowers and fine powders.

Condenser:-

Two designs of condensers are most popular in the essential oil industry. The older submerged coil type and the more modern shell and tube type.

Optimum design of condenser is critical because an undersized condenser will allow steam – oil vapors to escape uncondensed and over sizing the condenser means unnecessary extra capital cost.

Heat removal capacity of a condenser is expressed by the following equation:-

$$Q = U \times A \times T$$

Where

Q = Heat removal rate (BTU/hr)

U = Heat transfer co- - efficient of condenser expressed as; BTU/hr Ft20F)

A = Area available for heat transfer; (Sq ft.)

T = Log mean temperature difference between cooling water and condensate, (0F).

THE OIL SEPARATOR:-

- Oil separator has to perform the crucial function of separating the essential oil from the condensed steam.
- Generally oil is allowed to accumulate in the vessel, to be drawn off periodically, whereas the condensed water flows out continuously.
- A great variety of designs for separators are in use depending on oil density.
- There is an optimum temperature of the condensate at which the oil- water separation is most complete.
- Oil separator must have sufficient holding volume so that the entering oil – water mixture gets enough time to separate i.e. the residence time.
- Installation of a baffle barrier before the water outlet considerably improves separation of oil.

The Advantages and Disadvantages of Steam Distillation are as Follows:-

- The amount of steam and the quality of the steam can be controlled.
- Lower risk of thermal degredation as temperature generally not above 100 °C.
- Most widely used process for the extraction of essential oils on a large scale.
- Throughout the flavour and fragrance supply industry it is the standard method of extraction.
- There is a much higher capital requirement and with low-priced oils the pay back period can be over 10 years.
- Requires higher level of technical skill and fabrication and repairs and maintenance require a higher level of skill.

iii. PURIFICATION OF OIL

The insoluble particles present in the oil are removed by simple filtration method after mixing it with anhydrous sodium sulphate and keeping it overnight or for 4-5 hours. In case the colour of the oil changes due to rusting then it should be cleaned by steam rectification process.

8.0 STORAGE AND PACKING OF OIL

Lemongrass oil is stored in a dry place in the dark, in well-stoppered glass, or tin-lined or aluminum containers. Copper or iron containers dis-colour the oil. Light and air are detrimental. The oil undergoes gradual degradation on storage. When samples were stored under nitrogen in well-filled tightly closed vials, in a deep freeze, for 12 months, deterioration did not occur.

Lemongrass oil is also subject to easy oxidation resulting in lower citral content and lower bactericidal potency. Crude oils are more stable than refined oils. Over a 2 year period, crude oils were found to lose 50% of their citral content and purified oils 60%. It was found that betel leaves extract has the greatest protection against deterioration of citral after storing for one year. This may be. Attributed to the presence of chlorophyll in the extract. The only disadvantage with this antioxidant is that the oil gets the colour green. Sodium chloride 2%, which recorded 82.5% citral and the physical properties are within the ISI range, is also a good antioxidant. Pyrogallols (0.1, 0.5 and 1.0%) are effective in keeping up the citral content but they are confronted with the problem of toxic effects.

9.0 FRACTIONATION OF LEMONGRASS OIL

Fractionation is a separation process in which a certain quantity of a mixture (gas, solid, liquid, suspension or isotope) is divided during a phase transition, up in a number of smaller quantities (fractions) in which the composition varies according to a gradient. Fractions are collected based on differences in a specific property of the individual components. A common trait in fractionations is the need to find an optimum between the amount of fractions collected and the desired purity in each fraction. Fractionation makes it possible to isolate more than two components in a mixture in a single run. This property sets it apart from other separation techniques.

Fractionation is widely employed in many branches of science and technology. Mixtures of liquids and gases are separated by fractional distillation by difference in boiling point. Fractionation of components also takes place in column chromatography by a difference in affinity between stationary phase and the mobile phase. Infractional crystallization and fractional freezing, chemical substances are fractionated based on difference in solubility at a given temperature. In cell fractionation, cell components are separated by difference in mass.

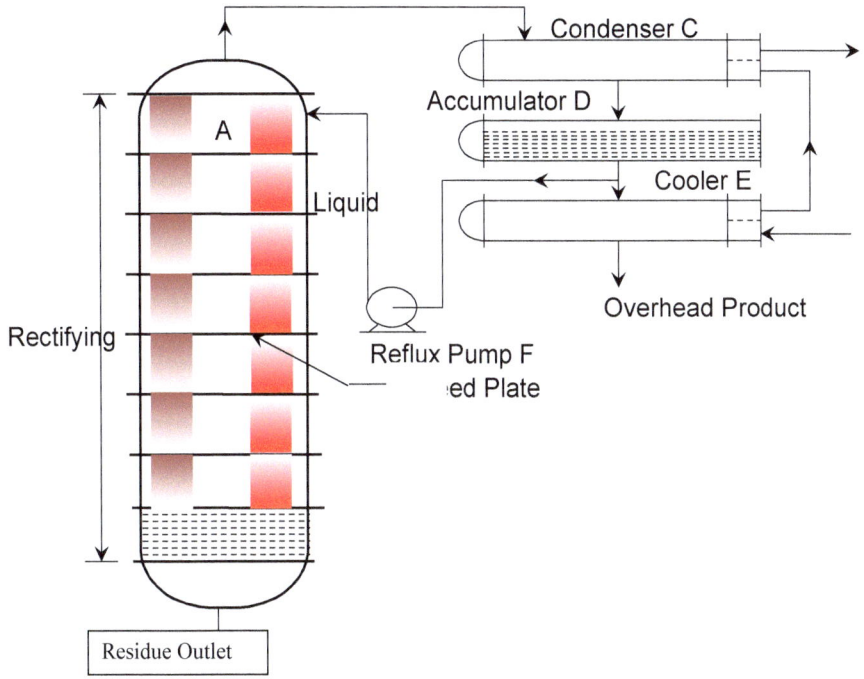

Figure 8 Diagram of Fractionation Column with Reflux and Distillate

i. FRACTIONATION AT LAB SCALE

The Fractionation unit at lab scale is shown in figure.
Isolation of citral from lemongrass oil was carried out by fractional distillation under reduced pressure with the following procedure:-

A measured quantity (500gm) of lemongrass oil was placed in a 3-neck round bottom flask in a fractional distillation set-up. This is equipped with a 3-neck round bottom flask which is attached to a column with metal packing materials and wrapped with electrical heating tape, a reflux divider, a condenser, multiple receivers and a trap. The high quality vacuum is also applied on the closed system to minimize the high temperature requirement in the system.

In the fractionation of Lemongrass oil firstly we start the heating at low temperature (almost 25 to 35 °C) which is fractionate the monoterpein part of lemongrass oil then we increase the temperature (almost 45 to 55) which is fractionate the Citral part of lemongrass oil in which present the Citral-a (Geranial) & Citral-b (neral).

Collection of the distillate or fraction was done every 10°C rise in temperature. When boiling slowed down, the heating rate was increased. Fractionation is continued until no more distillate is collected. About 3-4 fractions were collected for every distillation.

ii. Fractionation at Industrial Scale

9 250Liter Capacity Fractionation Column at FFDC Kannauj U.P.

Process Description

The fraction unit is shown in the figure (above).
1) Firstly we are clean the column by water vacuum for almost 15 to 20 minute to remove moisture content to the column, feed tank and all pipes which are closely connected to the column.
2) Then we charged the 180kg Lemongrass oil in the feed storage tank by water vacuum than disconnect the water vacuum pipe and apply the oil vacuum in the system.

3) In the feed storage tank have a stirrer rotated by 3HP electric motor to uniform mixing of feed.
4) In fractionation column have six distribution trays/stages which are connected by six pockets Sensor (temperature sensor).
5) Two condensers are connected to top part of the column which are have a vertical and horizontal.
6) The top outlet of condenser is connected to the trap which is followed by oil vacuum pipe.
7) After charging the lemongrass oil in feed tank we will start it's stirrer for uniform mixing and control the temperature of whole system.
8) The first fractionate of lemongrass oil is monoterpene which have lowest boiling point.
9) Then we collected the intermediate part, in which have monoterpene and citral.
10) Then we fractionate the citral which has major part of the lemongrass oil.
11) Then we collected the residue part by bottom of feed storage tank and the fractionate is collected by receiver.

Strategy of Fractionation Column Startups and Shutdowns

Column startup usually consists of the following steps:-
1. Commissioning. These operations clear the system of undesirable material and test it in preparation for introducing the chemicals. The demarcation between 'commissioning' and 'actual startup' is normally arbitrary, depending on the system and on subjective judgment.
2. Final elimination of undesirable materials.
3. Bringing the column to its normal operating pressure.
4. Column heating and cooling.
5. Introducing feed.
6. Starting up heating and cooling sources.
7. Bringing the column to the desired operating rates.

The sequence of steps performed at startup typically follows the above order, with some deviations.

Similarly, column shutdown usually consists of the following steps:-
1. Reducing column rates.
2. Shutting down heating and cooling sources.
3. Stopping feed.

4. Draining liquids.
5. Cooling or heating the column.
6. Bringing the column to atmospheric pressure.
7. Eliminating undesirable materials.
8. Preparing the column for opening to atmosphere.

The Condition necessary for good separation are:-
1. Comparatively large amount of liquid continually returning through the column.
2. Thorough mixing of liquid and vapour.
3. A large active surface of contact with the liquid and vapour.

REFLUX RATIO (R=L/D):-

L=Reflux Rate (kg/s)

D=Distillate Rate (kg/s)

Reflux is the liquid condensed from the rising vapor which returns to the pot flask. The Reflux ratio is the ratio between the boil up rate and the take-off rate. Or in other words, it is the ratio between the amount of reflux that goes back down the fractionation column and the amount of reflux that is collected in the receiver (distillate).

If 5 parts of the reflux go back down the distillation column and 1 part is collected as distillate then the reflux ratio is 5:1. In the case where all the reflux is collected as distillate the reflux ratio would be 0:1. If no distillate is collected then a reflux ratio is not assigned. Instead we call this "total reflux" or equilibration.

The higher the reflux ratio, the more vapor/liquid contact can occur in the distillation column. So higher reflux ratios usually mean higher purity of the distillate. It also means that the collection rate for the distillate will be slower.

PACKED DISTILLATION/FRACTIONATION COLUMN

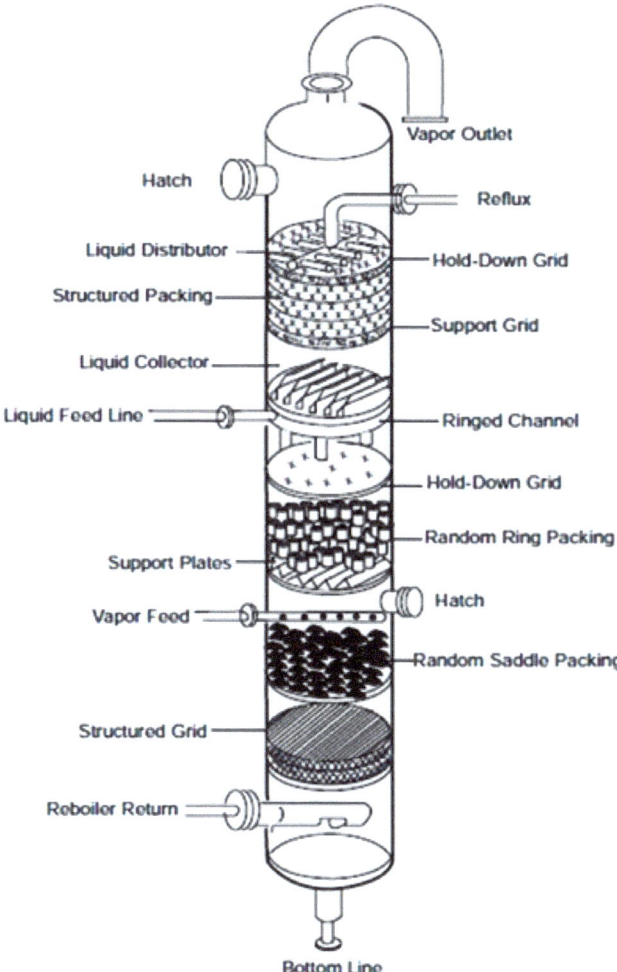

Figure 10 Packed Distillation Column

After fractionation of lemongrass oil the major part of lemongrass oil are following:-

Calculation based on 180 Kg lemongrass oil.

Chemical Family	Specific Components
1. Monoterpenes (10.8Kg)	[Myrcene + limonene] -6%
2. Intermediate (7.2 Kg)	[Citral + Monoterpene]-4%
3. Citral (126 kg)	[Geranial (Citral-a) + Neral (Citral-b)]-70%
4. Residue (34.2Kg)	19.22% Residue

MATERIAL BALANCE ON FRACTIONATION COLUMN

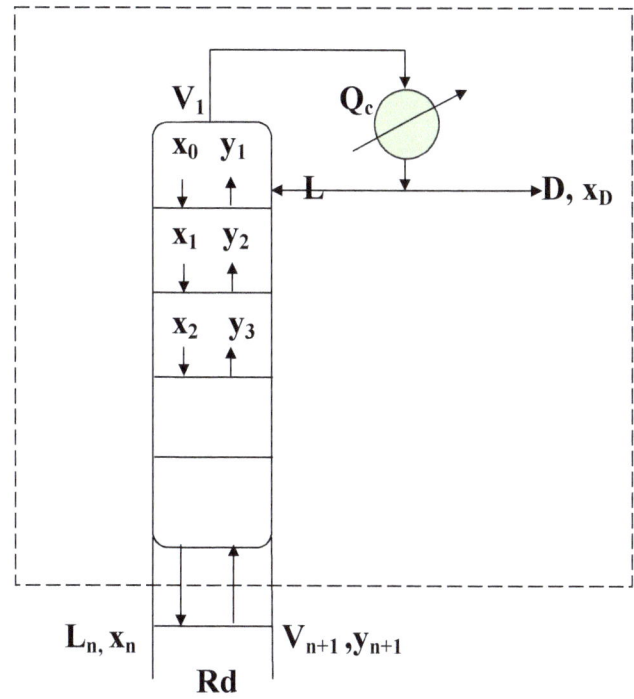

Figure 11 Material Balance on Fractionation Column

Material Balance on Fractionation Column:-

Figure shows the rectifying or enriching section which is the Fractionation tower section above the feed. The vapor from the top tray with a composition y_1 is condensed in the total condenser so that the resulting liquid is at the boiling point. Part of the liquid is taken out as the overhead product D and the remaining liquid is returned to the first tray with flow rate L.

Making a balance over the top part of the tower (the dashed-line section) on component A Gives:-

$$y_{n+1}V_{n+1} = x_n L_n + x_D D \quad \text{-----------(1)}$$

Since CMO (Constant Molar Overflow) is assumed
$\quad V_{n+1}$=V=Constant and L_n=L=Constant
Above equation can be written as-

$$y_{n+1} = x_n L/V + x_D D/V \quad \text{--------------------(2)}$$

From the total balance, V=L+D, D/V=1-L/V; in terms of Reflux ratio R=L/D;

$$\frac{L}{V} = \frac{L}{L+D} = \frac{L/D}{L/D + 1} = \frac{R}{R+1} \quad \text{-------------- (3)}$$

Eq. (2) becomes

$$y_{n+1} = \frac{R}{R+1}x_n + \frac{1}{R+1}x_D \quad \text{--------- (4)}$$

This equation which is the material balance for component A is called the operating line. Given x_D, x_F, R, and the equilibrium data, the equilibrium stages can be determined by successively solving the equilibrium relation and the material balance at each stage. Starting at x_D, y_1 is evaluated, for a total condenser $y_1 = x_D$. The liquid composition x_1 is then determined form the equilibrium curve at $y = y_1$. This is a dew point calculation with known y_1. Next y_2 is determined from the mass balance or operating line

$$y_2 = \frac{R}{R+1}x_1 + \frac{1}{R+1}x_D \quad \text{---------- (5)}$$

x_2 is again determined from the equilibrium curve and y 3 is determined from:-

$$y_3 = \frac{R}{R+1}x_2 + \frac{1}{R+1}x_D \quad \text{---------- (6)}$$

This process continues until $x_n < x_F$. The number of equilibrium stages in the rectifying section is n including the feed tray; n is the number of equilibrium calculations during the process.

Observation Table of Fractionation Column

S.No.	Feed Storage Tank Temperature (0C)	Vapor Temperature (0C)	Reflux Ratio (L/D=R)	Vacuum Of the system (mm Hg)	Fractionate or Distillate
1.	122 °C	68 °C	30	5mm Hg	Monoterpene
2.	125 °C	77 °C	15	5mm Hg	Intermediate
3.	139 °C	84 °C	15	5mm Hg	Citral

Material Balance on Processed Material (Lemongrass Oil)

A. For fractionation of monoterpene:-

Input = Output + Accumulation + Disappearance
We are assuming process is steady state
so,

Accumulation=0

(180 kg Lemongrass Oil) = (10.8 kg Monoterpene) + (168.7kg Residue) + (X_1 kg process loss)]
X_1=(0.4 kg process loss)

B. For fractionation of Intermediate:-

(168.7 kg Residue) = (7.2 kg Intermediate) + (161.3 kg Residue) + [X_2 kg Process loss)
X_2=(0.2 kg Process loss)

C. For Fractionation of Citral:-

(161.3 kg Residue) = (126 kg Citral) + (34.6 kg Residue) + (X_3 kg Process loss)
X_3=(0.7 kg Process loss)

Total Process Loss=X_1+X_2+X_3
Total Process Loss=0.4+0.2+0.7=1.3kg

Operating Line for the Rectifying Section of the Fractionator

$$y_1 = \frac{R}{R+1} x_0 + \frac{1}{R+1} x_D$$

1) For Monoterpene Fractionation:-
Given:-
R=L/D=30

$$y_1 = \frac{30}{30+1} * x_0 + \frac{1}{30+1} * x_D$$
$$y_2 = 0.92307 * x_0 + 0.07692 * x_D$$

2) **For Citral:-**

$$R = L/D = 5$$
$$y_1 = \frac{5}{5+1} * x_0 + \frac{1}{5+1} * x_D$$
$$y_2 = 0.833 * x_0 + 0.167 * x_D$$

Energy Balance of Condenser of Fractionating Column

Figure 12 Glass Fractionation Unit (Capacity-25 liter) in Process Division at FFDC, Kannauj

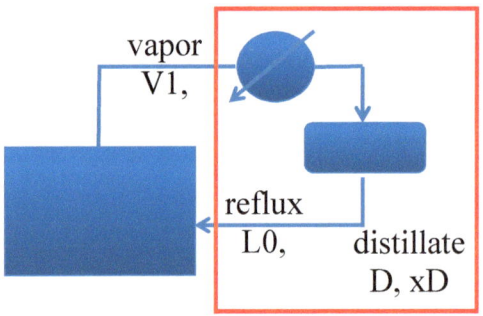

Figure 13 Energy Balance on Condenser of Fractionation Column

S.No.	Feed Storage Tank Temperature (0C)	Vapor Temperature (0C)	Reflux Ratio ($L_0/D=R$)	Pressure Of the system	Fractionate or Distillate
1.	250 °C	225 °C	30	1 atm	Monoterpene
2.	255 °C	228 °C	15	1 atm	Intermediate
3.	260 °C	231 °C	15	1 atm	Citral

Energy balance

$$V_1H_1 + Q_C = (D + L_0)h_D = V_1h_D$$

$$Q_C = V_1(h_D - H_1)$$

Given:-

$V1 = 14$ Kg   (Total Mass of Vapor at 231 °C)

$h_D = 26.5$ KJ/mol = 0.17374 KJ/kg   (Enthalpy of distillate citral at 100 °C temperature)

$H_1 = 48.6$ KJ/mol = 0.31863 KJ/kg   (Enthalpy of Vapor (citral) at 231 °C temperature)

So

$Q_C = 14$kg $*(26.5-48.6)$ KJ / mol
$Q_C = 14$kg$*[-22.1*$KJ / (152.53kg)]
$Q_C = -2.02845$KJ
$Q_C = -2028.45$ Joule

Matlab Program for Energy Balance on Condenser

```
function Qc =Condenser (V1, hD, H1);
% V1=Total Mass of Vapor at 231  0C;
% hD=Enthalpy of distillate citral at 100  0C temperature;
% H1=Enthalpy of Vapor (citral) at 231  0C temperature;
V1=input('Enter the value of V1 in kg:');
hD=input('Enter the value of hD in kj/kg:');
H1=input('Enter the value of H1 in kj/kg:');
Qc=V1*(hD-H1);
end
```
Run Condenser.m (F5)

Enter the value of V1 in kg:14
Enter the value of hD in kj/kg:0.17374
Enter the value of H1 in kj/kg: 0.31863
ans =

-2.0285KJ

Requirement of Water in Condenser for Cooling

We know that $Q = mC_p\Delta T$
Given:-

    m=Total Kg Water Required For Condensation
    Q= -2028.45 Joule (Heat Required for Condensation)
    T_1=25 °C (Inlet Temperature of Water in Condenser)
    T_2=35 °C (Outlet Temperature of Water in Condenser)
    Cp= 4.18 Joule/kg °C (Specific Heat of Water)

-2028.45 J= m*4.18* (25-35) J/Kg

m = **48.5275** kg Water Required for Condensation

Matlab Program

```
function m =Condenser1 (Q, Cp, T1, T2);
% Qc=Removable heat of the vapor by cooling water;
% m=Total Kg Water Required For Condensation;
% T1=25  OC (Inlet Temperature of Water in Condenser);
% T2=35  OC (Outlet Temperature of Water in Condenser);
% Cp=4.18 Joule/kg OC (Specific Heat of Water;
Q=input('Enter the value of Q in Joule:');
T1=input('Enter the value of T1 in OC:');
T2=input('Enter the value of T2 in OC:');
Cp=4.18;%Joule/kg OC
m=Q/(Cp*(T1-T2));
end
```

Run Condenser1.m (F5)
Enter the value of Q in Joule: -2028.45
Enter the value of T1 in 0C: 25
Enter the value of T2 in 0C: 35

ans =

48.5275

10.0 TROUBLESHOOTING

i. Technique for Preventing Tray Damage Due to Excessive Liquid Level

1. Pump liquid out rather than attempt to boil it off if liquid level is excessive.
2. Construct the bottom seal pan to be particularly strong.
3. Construct the bottom 25 percent of trays for extra mechanical strength.
4. Provide a liquid level differential pressure measurement for the bottom 25 percent of the column.
5. Provide facilities for easy diversion of bottom liquid to either the feed tank or storage so that liquid level can be readily reduced.
6. Ensure smooth and stable automatic control of boil-up to the tower.

ii. Sources and Effects of Water Problems

Water can causes severe operational problems in services which are not meant to handle it. One refiner stated that 99 % of his fractionators upsets are due to water; other refiners agreed with this evaluation. Most problems occur when the column separates water-insoluble materials such as hydrocarbons. The main adverse effects of water in such services are pressure surges, flooding, cycling, corrosion, hydrates, and off-spec products.

Typical sources of water in this type of services are:-

1. The feed stream. Water may be present in feed storage tank, may enter the feed from a leaking heat exchanger, or may not be entirely removed in an upstream removal facility.
2. Un-drained water in a stripping a steam line.
3. Chemical reaction. A condensation reaction forming water may occur between organic chemicals, either in upstream equipment or in the column.
4. A leaking heat exchanger (e.g. reboiler, condenser).
5. A prestart-up wash, leak-test, or steam water operation.
6. The previous campaign runs through the column.

7. Condensate formed in previous operations. It could have remained trapped in pipelines or pockets inside the equipments.

8. A water-containing stream that found its way to the column (a typical example is water discharged from the de-salter safety valve and ends up in a refinery crude column).

11.0 Lemongrass Oil Processing Economics

Processing Lot = 180 Kg
Lemongrass Oil Cost:-
[(180kg) * (1000Rs/kg)] = 1, 80, 000 /- **INR**

Processing Charges :-(Electricity Charges + Diesel Consumption) INR

1.Cooling Tower (7.5 HP for 9 hrs)	201.42/-
2.Water Vacuum Pump (15HP for 1 hrs)	44.76/-
3.Oil vacuum pump (7.5 HP for 6 hrs)	134.28/-
4.Thermic fluid heater pump(10HP for 8 hrs)	238.72/-
5.Gear Box-Stirrer (3 HP for 8 hrs)	71.616/-
6.Thermic Fluid pump & blower (3 HP for 6 hrs)	53.712/-
7.Diesel Consumption(unit work 6hrs & consumed 9lit/hr)	3240/-
Total	**3984.508/-**

Manpower Charges: - (all price rate in INR)

1.Supervisor-1 (600*1)	600/-
2.Labour-2(200*2)	400/-
Total	**1000/-**

Maintenance Charges: - (all price rate in INR)

1.Fractionation Column	55/-
2.Plant Building Charge	133/-
3.Thermic Fluid Heater	33/-
4.Cooling Tower	10/-
5.Water Vacuum	5/-
6.Oil Vacuum	5/-
Total	**241/-**

Total Investment in 1 batch of 180 kg lemongrass oil: - (price rate in INR)

1.Total Processing Charges (3984.508+1000+241)	5225.508/-
2.Lemongrass Oil Cost (180 kg)	1,80,000/-
Total	**1,85225.508/-**

Total Return:-
Products and its return :-(all price rate in INR)

1. Monoterpene (10.8Kg*350)	2700/-
2. Intermediate Part (7.2Kg*300)	2160/-
3. Citral (126*1800)	2,01,600/-
4. Residue (34.2*250)	8550/-
Total	**2,15,010/-**

Total Margin = 2,15,010– 1, 85,225.508 = 29,784.492 INR /180 kg oil
*The total profit on 1 kg lemongrass oil fractionation is 165.4694 **INR**.*

12.0 Uses of Lemongrass Oil

Lemongrass oil can be used either as an oil in its own right or as a route to the isolation of its derivative citral. Citral can be further processed to isolate a group of chemicals known as the ionones, which possess a violet like fragrance and are important components in many articles of perfumery. A further processing of these ionones makes possible the manufacture of vitamins, notably vitamin A.

i. PERFUMERY

The oil is currently used more as oil in its own right in cheap fragrance work. Its many applications include aerosol deodorants, floor polishes, household detergents and other domestic and industrial products requiring a fresh, pleasant fragrance or a mask for the unpleasant odours of certain active ingredients. The odour of the oil resembles that of Verbena oil. 98-100% citral gives a lemon odour to soap after one month, but the odour fades within 6 months. The soap is also liable to yellow slightly. Ionones are a group of very important synthetic aromatics possessing a strong and lasting violet odour.

The following values for the concentration of lemongrass oil in products, when used as a perfume.

	Soap	Detergents	Creams Lotions	Perfume
Usual	0.02 %	0.002 %	0.003 %	0.08 %
Maximum	0.25 %	0.025 %	0.02 %	0.7 %

ii. INSECT REPELLENT PROPERTIES

As an insect repellent it shows great promise and can be combined with the synthetic organic insecticides DDT and BHC.

iii. GERMICIDAL AND BACTERICID

Lemongrass oil has good antibacterial properties. In preparing a disinfectant solution with lemongrass oil, the least amount of solvent must be used.

It gives a clean smell to hospital rooms and closets. It is a constituent of dermo-cosmetics.

iv. THERAPEUTIC USAGE

Lemongrass is said to be a stimulant, diaphoretic and anti-spasmodic and its oil is carminative and tonic. Externally it is rubefacient.

According to the Indian Materia Medica, lemon grass oil is useful as a carminative in flatulent and spasmodic affections of the bowels, colic, gastric irritability, and is of great value in cholera with obstinate vomiting. The dose is from 3 to 6 drops on a piece of loaf sugar or in emulsion.

An infusion or decoction of 4 ozs. of the grass to 1 pint of boiling water is an excellent stomachic to children; with ginger, sugar, cinnamon, it is given as disphoretic in fevers, and given with black pepper it is useful in dropsical conditions caused by chronic malaria.

Mixed with equal quantity of pure coconut oil it makes an excellent embrocation or liniment for lumbago, chronic rheumatism, neuralgia, sprains and other painful affections; it is also a good application for ringworm.

A tea of lemongrass leaves is used in Brazil and other Third World countries as a popular remedy for various nervous and gastrointestinal disturbances.

West Indian lemongrass leaves are used in Cuba, as an anti-hypertensive and anti-inflammatory folk medicine. A 10% or 20% decoction of leaves showed some dose-related hypertension effects when given intravenously and some weak diuretic and anti-inflammatory effect when given orally.

Herbal tea prepared from Cymbopogon citratus leaves showed alleged CNS - depressant effects, is not toxic but also lacks hypnotic or anxiolytic properties.

v. OTHER USES

Lemongrass oil has been reported for several other minor uses. It is said that the oil can be used to improve the flavour of some fish and can be used to flavour wines, sauces, confectionery, spices and tea leaves.

The safe usage level in food products is:-

Non-alcoholic Beverages	4.4 ppm

Ice cream, ices etc.	9.2 ppm
Candy and baked goods	38 ppm
Gelatine and puddings	290 ppm

13.0   REFERENCES

- Abegaz, B. And Yohannes, P.G. Constituents Of The Essential Oil Of Ethiopian Cymbopogon Citratus Stapf J. Natural Products 46(3), P.424-426 (1983).
- Absar, Ahmad; Janardhanan, K.K. and Verma H.N. In vitro growth of Balansia sclerotica (Pat.)Hohn, an endophyte causing grassy-shoot disease of lemongrass. Kavaka 19(l-2), p.48-52 (1994).
- Ansari, S.H.; Bhatnagar, J.K. and Qadry, J.S., Thin layer and gas chromatographic analysis of lemongrass oil. Indian J. Nat. Prod. 2(2), p.3-7 (1986).
- Secondini, Olindo "Handbook of Perfumer and Flavors", Chemical Publishing Co. Inc. New York 1990.
- Atal, C.K. (Ed.) and Kapur, B.M. (Ed.) "Cultivation and utilization of aromatic plants", Council of Sci. & Indust. Res., India. 1982.
- Perry, R. H., and C. H. Chilton (eds.), Chemical Engineers' Handbook, 5th ed., McGraw-Hill, New York (1973).
- Seader, J. D. , and E. J. Henley, Separation Process Principles, 2nd ed.,Wiley, New York (2006).
- Treybal, R. E., Mass-Transfer Operations, 3rd ed., McGraw-Hill, New York (1980).
- McCabe, W. L., J. C. Smith, and P. Harriott, Unit Operations of Chemical Engineering, 7th ed., McGraw-Hill, New York (2005).

www.ingramcontent.com/pod-product-compliance
Lightning Source LLC
Chambersburg PA
CBHW040333220526
45473CB00009B/2665